The Theory of Spiral Stabilization

Continuation of the Cosmic Twist and Spiral Model

THEORY OF SPIRAL STABILIZATION

First edition. March 14, 2026.

ISBN: 979-8233445125

Written by MATE ISTVAN-MATYAS.

Written by: Máté István-Mátyás

Introduction

Between Order and Chaos

The world we live in is more complex than any system man has ever built. Nature, society, living organisms, and technology all function as vast networks in which countless elements are interconnected. These systems are in constant motion: they exchange energy, transmit information, and continuously adapt to their environment.

Yet, there is a common question that lies behind every complex system: **what holds them together?**

What prevents a sophisticated system from collapsing under its own complexity? What is the mechanism that allows living organisms, ecosystems, nervous systems, or even human-made technological networks to operate stably over long periods of time?

Throughout history, people have often thought of stability as a kind of static state—a point of rest where all change ceases. Nature, however, shows us something entirely different. Most stable systems are not motionless. On the contrary: **they operate in rhythm.**

The heart beats. The brain generates waves. The seasons alternate in cycles. Stars pulsate. Galaxies rotate in spirals.

Stability, therefore, is not necessarily a sign of permanence. Rather, it is a **dynamic equilibrium** that remains in continuous motion.

This realization led to the concept that we call in this book the **Spiral Stabilization Principle (SSP)**. The SSP starts from a simple but powerful idea: the stability of a complex system can be maintained if the system is capable of dampening excessive fluctuations while amplifying signals that are too weak. In other words, the system continuously corrects its own operation and keeps its dynamics within a stable range.

This mechanism does not stop motion. On the contrary: **it enables the system to function in rhythm.**

The purpose of this book is not to prove a new cosmological theory or to provide a single explanation for every phenomenon in nature. Rather, it is a thought experiment: could there exist a general stabilization principle that helps us understand how complex systems remain operational?

In the following chapters, we will examine this idea from several perspectives. First, we approach it through patterns observable in nature, then we present the mathematical form of the Spiral Stabilization Principle, and finally, we explore how this principle can be applied to networks, artificial intelligence, and other complex systems.

It is possible that the SSP will be just one useful model among many. But it is also conceivable that it offers a perspective that helps us look at the question of stability in a different way.

Because perhaps the order of the world does not lie in stillness. Perhaps order itself is **motion**.

Chapter 1

The Edge of Chaos

One of the most important realizations of modern science is that most natural systems are neither perfectly stable nor entirely chaotic. The majority of systems operating in nature exist in a special state often referred to as the **edge of chaos**.

This is the state where a system has not yet collapsed into disordered noise, but is no longer rigid and unchanging. It is here that the delicate balance appears, enabling adaptation, evolution, and the formation of complex structures.

If a system is too rigid, it cannot adapt to changes. A completely stable, static system easily becomes fragile because any unexpected disturbance disrupts its balance. At the same time, an excessively chaotic system cannot survive either, as the patterns and connections within it are constantly falling apart.

Most functional systems are located somewhere between the two.

This state is particularly observable in living systems. The human brain, for instance, is in constant activity, yet it is capable of stable operation. The vast network of neurons generates continuous electrical activity, but this activity is not entirely random. Various rhythms emerge in the brain's functioning: alpha waves, beta waves, and other oscillations that reflect the system's dynamic stability.

A similar phenomenon can be observed in ecosystems. A forest or an ocean's wildlife is in constant flux. Populations rise and fall, species compete with one another, and the environment is perpetually modifying. Yet, over long periods, a relatively stable state is maintained. The system is not motionless; rather, it **continuously corrects its own operation**.

Nature, therefore, often does not seek total stability, but rather a state where change and stability are present simultaneously.

This phenomenon also appears in physics and mathematics. In many non-linear systems, there exist so-called **attractors** toward which the state of the system converges over time. These attractors are not always point-like; they often appear in the form of cycles, spiral trajectories, or more complex patterns.

Cyclic attractors are particularly interesting because the system does not reach a point of rest, but instead settles into a stable rhythm. This means that stability is not the result of motionlessness, but of a **recurring dynamic**.

Nature offers numerous examples of this operation. The rhythm of a heartbeat is a stable biological cycle. The orbit of planets is a periodic motion maintained by gravity. Some stars are known as pulsating variables, whose brightness changes in regular cycles. Even in the structure of galaxies, spiral patterns appear.

These phenomena suggest that stable systems are often **rhythmic or cyclic in nature**.

The concept of the Spiral Stabilization Principle (SSP) stems from this observation. If stable systems often exhibit cyclic dynamics, then it is conceivable that stability originates not from a static state, but from a specific feedback mechanism.

Such a mechanism may be capable of keeping a system within a stable range while allowing for continuous motion and adaptation.

In other words, stability does not necessarily mean that the system stops. In many cases, it means that the system **finds its own rhythm**.

In the next chapter, we will examine a geometric pattern that appears with surprising frequency in nature, and which may provide a key to better understanding this dynamic stability.

This pattern is the **spiral**.

Chapter 2

The Spiral is Everywhere

If one observes nature attentively, a peculiar phenomenon soon becomes apparent: certain forms and patterns reappear time and again in systems that are entirely independent of one another. Among these, one of the most common and mysterious forms is the **spiral**.

Spiral structures can be found in both the largest and the smallest structures of the universe. Astronomers are well-acquainted with spiral galaxies, in which billions of stars are arranged into a vast, rotating pattern. The same geometry appears in the rotating cloud arms of cyclones and hurricanes, in the movement of swirling water, or even in the pattern of ripples created by throwing a simple pebble into a pond.

The spiral, however, does not only appear in cosmic and physical systems. In the living world, it is perhaps even more frequent. The arrangement of leaves on plants, the pattern of sunflower seeds, the growth of seashells, or the houses of snails all follow a spiral structure. One of the best-known examples in biology is the double helix of DNA, which carries the genetic information of living organisms.

These examples suggest that the spiral is not merely an aesthetic form, but **a natural mode of organization**.

But why does it appear so frequently?

One special property of the spiral is that it expresses **growth and stability** simultaneously. A spiral form is capable of continuously expanding or contracting while preserving its fundamental structure. This is particularly useful in systems that must adapt to changing conditions while maintaining internal order.

Mathematically, a spiral can often be described as a motion in which a point moves outward or inward while continuously rotating around a center. This combination—rotation and expansion—can create extremely stable trajectories. In many dynamic systems, motion naturally converges toward spiral paths when a system moves toward equilibrium.

This phenomenon is well-known in mathematics and physical systems. When a system approaches a stable state, it often does not move toward it in a straight line, but approaches it spirally. The trajectory describes increasingly smaller circles while the system's energy gradually decreases.

This type of motion is a kind of **dynamic ordering**.

Therefore, the spiral is not only a form but also a **process**. It is a process that allows systems to adapt, correct their fluctuations, and ultimately create a stable pattern.

This intuition also lies behind the Spiral Stabilization Principle. If many stable systems in nature exhibit spiral or cyclic motion, then it is conceivable that one of the keys to stability is **spiral dynamics**.

In this framework, the system does not strive toward a single point but settles into a stable pattern of motion. Stability, therefore, is not necessarily a static state, but a motion capable of persisting over long periods of time.

In the next chapter, we will approach this intuition from a specific question: what happens to a system when excessive fluctuations appear within it, or when the system's energy drops too low? How does a system react to its own instability?

These questions lead us to the fundamental idea of the SSP: the **mechanism of spiral stabilization**.

Chapter 3

The Search for Equilibrium

In every complex system, there is a fundamental tension between two opposing tendencies. One direction leads toward excessive activity: the system's energy increases, movements amplify, and fluctuations become increasingly large. The other direction is the opposite: the system's energy decreases, the dynamics slow down, and the system may eventually enter a passive state.

These two extremes appear in many forms in nature. In an ecosystem, the overly rapid growth of a population can lead to resource depletion and collapse. If, however, the population becomes too small, it is unable to sustain itself. In economic systems, excessive growth can induce the formation of bubbles, while too great a decline can lead to recession. In the functioning of the nervous system, excessive activity can cause epileptic seizures, while too low activity can lead to system failure.

These examples all suggest that for stable operation, it is not enough for a system to simply remain at rest. Stability means, much more so, that the system is capable of continuously correcting its own state.

A stable system, therefore, is not passive. On the contrary: it actively reacts to its own changes.

Let us imagine a simple example. If a system shows an excessively large fluctuation, there is a need for a mechanism that dampens this fluctuation. If, however, the system enters an excessively weak or low-energy state, an influence is needed that amplifies its operation once again.

In other words, stability is often created from the balance of two types of feedback:

- A **negative feedback** that reduces excessive

fluctuations;
- and a **positive feedback** that strengthens states that are too weak.

This dual mechanism is observable in many natural systems. Living organisms, for instance, continuously regulate their internal processes. Body temperature, blood sugar levels, or hormonal regulation are all systems in which the organism constantly corrects deviations.

In engineering systems, this same idea appears in control theory. A well-designed system senses its own state and applies feedback that guides its operation back into a stable range.

The Spiral Stabilization Principle generalizes this fundamental idea. Instead of searching for a single stable point, we can imagine a mechanism that keeps the system within a **stable range**. Within this range, the system is capable of moving, changing, and adapting, but the fluctuations do not grow without limit.

This idea is particularly important in large, networked systems. In such systems, it is often impossible to maintain a single, precise state of equilibrium. Stable operation arises much more from the fact that the system's dynamics remain within a certain range.

The Spiral Stabilization Principle, therefore, does not assume a rigid equilibrium. Rather, it describes a dynamic system capable of moving within a stable zone.

In the next chapter, we will cast this concept into mathematical form and present the simple model that provides the foundation for the Spiral Stabilization Principle.

Chapter 4

The Mathematics of Spiral Stabilization

In the previous chapters, we examined intuitively how stability might function in complex systems. We have seen that many natural systems do not approach a single static equilibrium point, but rather remain stable within a dynamic range. The next step is to formulate this concept in a mathematical form.

Most dynamic systems can be described by a simple equation that shows how the state of the system changes over time:

$$\dot{x} = f(x)$$

Here, x is the current state of the system, and $f(x)$ is the rule that determines how this state changes. In many cases, this dynamics alone does not guarantee stability. If the system produces fluctuations that are too large, the value of the state can increase without limit or, conversely, drop to zero.

The Spiral Stabilization Principle addresses this problem with a simple idea: we add a stabilizing feedback to the system's operation.

The new equation of the system can be written as follows:

$$\dot{x} = f(x) + u(x)$$

In this expression, $u(x)$ is a regulatory term that depends on the current state of the system. Its task is to correct fluctuations that are either too large or too small.

One of the simplest forms of Spiral Stabilization appears when we interpret the state of the system as a radius in an abstract space. Let us denote this quantity by r. The goal of the stabilizing term is to keep this radial value within a given range.

One possible form of the stabilizing function is the following:

$$u(r) = -\alpha \cdot sp_\kappa(r - r_h) + \beta \cdot sp_\kappa(r_\ell - r)$$

In this equation, r_h and r_ℓ denote two limit values that define the stable range. The sp_κ function is a smooth transition function that creates a gradually strengthening feedback.

The operation of the equation is easy to understand intuitively.

If the state of the system exceeds the upper limit ($r > r_h$), the negative term is activated and dampens further growth. If, however, the state of the system falls into the too-low range ($r < r_\ell$), the positive term comes into play, strengthening the system's dynamics.

The combined effect of these two mechanisms leads to the formation of a stable range:

$$r \in [r_\ell, r_h]$$

Within this range, the system can move freely, but the fluctuations do not grow indefinitely. Thus, the system's dynamics are not organized around a single point but remain within a stable zone.

This type of stability is particularly important in large, noisy systems. Real-world systems rarely operate with perfect precision. Noise, external disturbances, or unexpected changes are always present. A system that is too rigid can easily become unstable under such conditions. A stable range, however, allows the system to adapt flexibly.

The mathematical model of Spiral Stabilization thus provides a framework in which the system is not tied to a single equilibrium point but is capable of operating within a stable range.

In the next chapter, we will examine how a special dynamic develops within this range: the system's own rhythm, which can be considered one of the most important phenomena of Spiral Stabilization. This phenomenon is the so-called **stable cycle**, or otherwise known as the system's "heartbeat."

Chapter 5

The Heartbeat of the System

In the previous chapter, we saw that the Spiral Stabilization Principle creates a stable range in which the system's dynamics can function safely. This range prevents the system from entering extreme states. However, stability does not mean that the system becomes motionless.

In fact, within the stable range, a special phenomenon often emerges: **rhythm**.

This rhythm originates from the system's own natural dynamics. When the stabilizing mechanism keeps fluctuations in check, the system does not converge toward a single point but instead adopts a recurring pattern of motion. Mathematically, this state is called a **limit cycle**.

A limit cycle is a closed trajectory in the state space toward which the system's motion gravitates over time. The system does not stop on this path; rather, it traverses it continuously. This continuous motion provides the system's stable rhythm.

In the Spiral Stabilization model, this phenomenon appears when the system's radius approaches a stable value:

$$r \rightarrow r^*$$

This value is located within the stable range, and the system's motion organizes itself into a trajectory that maintains this radial value.

The result is a stable oscillation.

This phenomenon is extremely common in nature. The heart, for example, does not function in a state of rest, but maintains circulation through continuous pulsation. In the nervous system, the activity of neurons is organized in the form of waves. In astronomy, many stars are known whose brightness changes in regular cycles.

In these systems, stability does not stem from stillness, but from the system finding its own rhythm.

The Spiral Stabilization Principle interprets this phenomenon as a natural consequence. When both excessively large and excessively small fluctuations are corrected through feedback, the system's dynamics often settle into a stable cycle.

Thus, the stable range not only provides protection against instability but also enables the system to develop an **organized rhythm**.

This idea has important implications when we examine not just a single system, but a network of systems. If several interconnected systems all possess such a stable rhythm, the question arises: are these systems capable of aligning with one another?

In other words, can a **common rhythm** emerge within a network?

In the next chapter, we will examine this question and demonstrate how Spiral Stabilization can lead to network synchronization.

Chapter 6

Network Sync – When Systems Breathe Together

In the previous chapter, we examined how a stable rhythm emerges in the operation of a single system. We have seen that Spiral Stabilization does not create a static equilibrium, but a stable cycle in which the system moves continuously.

Nature, however, rarely operates with isolated systems. Most systems are connected to others. Cells communicate with each other, neurons form networks, planets interact gravitationally, and humans are part of social networks.

When several systems operating rhythmically come into contact with one another, a new phenomenon appears: **synchronization.**

Synchronization means that the rhythms of separate systems align with each other. They will not necessarily become completely identical, but a common pattern begins to emerge.

Mathematically, this phenomenon is often described by a network of oscillators. In a simple model, every system has its own phase:

$$\theta_i$$

where i is the index of the given system. The interaction of the systems can be described by the following equation:

$$\dot{\theta_i} = \omega_i + \sum_{j} K_{ij}\sin(\theta_j - \theta_i)$$

In this equation:

- ω_i is the natural frequency of the individual systems;
- K_{ij} is the strength of the connection between the systems;
- and the sine term regulates the phase difference.

When the connections are strong enough, the system moves toward a **common synchronous state.**

The Spiral Stabilization Principle plays a special role in this process. Since every system oscillates within a stable range, the fluctuations cannot become too large. This prevents a single system from making the entire network unstable.

In other words, the stable range **dampens noise** while the rhythm is preserved. This mechanism allows synchronization to develop gradually within the network.

This phenomenon can be observed in many places in nature. Fireflies, for example, start to flash in unison after a while. Heart cells are capable of spontaneously coordinating their activity. Neurons in the brain organize into rhythmic waves. Even pendulum clocks can fall into sync if they are hung on the same wall.

All these phenomena show that synchronization is not an exception, but a natural property of networked systems.

The Spiral Stabilization Principle complements this by providing a stable range for the systems. Within this range, the rhythm can persist while the network gradually becomes more ordered.

This idea becomes particularly important when the number of systems is very large. The operation of a network consisting of millions or billions of elements can no longer be simply described by a single equation.

Such systems include:

- the Internet;
- economic networks;
- ecological systems;
- or even human society itself.

These systems often appear chaotic because of the countless interactions occurring simultaneously. However, the Spiral

Stabilization Principle suggests that a stabilizing mechanism can exist even behind chaotic behavior.

If this mechanism truly exists, a deeper question arises.

Is it possible that stability and rhythm are not just properties of networks, but **the fundamental operating principles of the universe?**

In the next chapter, we will examine this question and attempt to place the Spiral Stabilization Principle within the context of the great systems of nature.

Chapter 7

The Universe as a Stabilized Network

When the Spiral Stabilization Principle was first formulated, we were primarily looking for a new model to understand the functioning of complex systems. However, as the investigations went deeper, a surprising thought began to take shape.

It is possible that the stabilization mechanism we observe in networks and systems is not merely a local phenomenon. It is conceivable that the universe itself operates according to similar principles.

According to modern cosmology, the universe does not consist of a homogeneous distribution of matter but of an extremely complex structure. Galaxies are not positioned randomly; they are organized into vast filaments and nodes. Together, these structures form what researchers call the **cosmic web**.

In this web, galaxies and galaxy clusters appear as nodes, while the filaments stretching between them mark the flow paths of matter and energy.

If we examine this structure from the perspective of network theory, it becomes apparent that the large-scale structure of the universe bears a surprising resemblance to other complex systems. We find similar patterns, for example, in:

- the network of neurons in the brain;
- the topology of the Internet;
- or the relational structures of ecological systems.

These systems often exhibit the same statistical properties as **scale-free networks**. In such networks, a few nodes have an extremely high number of connections, while most nodes have only a few.

This structure makes the system particularly resilient. If a few random nodes drop out, the network remains operational. At the same time, energy, information, or matter can flow efficiently along the network.

The Spiral Stabilization Principle takes on a new interpretation in this context. The stable range can ensure not only the operation of a single system but can also form the basis for the stability of an entire network.

If the universe indeed functions as a network structure, its stability is likely not the consequence of a single physical law but emerges from the balance of many interactions. In this framework, information and noise are in constant interaction.

Light—which carries information—is one of the fundamental communication channels of the universe. Gravity and other interactions, meanwhile, direct the formation of structures.

If we interpret the universe in this way, stability is not a passive state but an active process. Systems continuously react to one another, adapt to changes, and maintain equilibrium through feedbacks.

This idea is close to the concept that the universe is not a static machine but a **dynamic, self-organizing system**.

In this picture, the Spiral Stabilization Principle may describe a mechanism that allows complex systems—be they cells, societies, or galaxies—to operate stably over long periods of time.

But if this is truly the case, a even deeper question arises. What happens when the balance between information and noise is disrupted?

In the next chapter, we will examine this question and look at how complex systems react to instability.

Chapter 8

When the System Tips – Instability and the Birth of a New Order

In the previous chapter, we explored how stability emerges in complex networks and raised the possibility that the universe itself functions as a dynamic, stabilized network. This picture, however, is only one half of the story.

Complex systems do not always remain in a stable state. From time to time, changes occur within them that tip them out of equilibrium. These states often appear chaotic, but experience shows that instability does not necessarily mean collapse. In many cases, it is precisely instability that creates the possibility for a new order to emerge.

Physics and mathematical system modeling know many examples of this. When the parameters of a system change gradually, at a certain point, the previous stable state ceases to exist. This phenomenon is called **bifurcation**.

During a bifurcation, the system enters a new mode of operation. Instead of the previous stable point, for example, two new stable states may appear, or the system may enter cyclic behavior.

This process is frequently observed in nature. The flow of fluids, for instance, becomes turbulent above a certain velocity. In living systems, cell differentiation creates new functions. In social systems, entirely new structures emerge from time to time.

From the perspective of the Spiral Stabilization Principle, these processes are particularly interesting. The stable range is not a rigid boundary, but a region within which the system is capable of adapting to changes. However, if the system strays too far from this range, the stabilizing mechanism can no longer pull it back into equilibrium.

When this happens, two things can occur. One possibility is collapse: the system loses its organization, and the dynamics become disordered. The other possibility, however, is much more interesting: **the system finds a new stable range.**

This process is often spiral in nature. The system swings out of equilibrium and then returns to a new trajectory. The new state is not identical to the old one, but it may be more stable under the new conditions.

This phenomenon may be one of the fundamental mechanisms of development. In biology, evolution works in a similar way. Environmental changes create instability, which forces new adaptation strategies. The development of technological systems also often occurs through such leaps.

In this context, the Spiral Stabilization Principle articulates an important thought: stability and change are not opposites. In fact, they are two sides of the same process. Stability allows a system to function for a long time, while instability creates the opportunity for the system to find new modes of operation.

This dynamic is a kind of **balance between order and chaos.**

Many researchers believe that life and intelligence appear precisely in this border zone. If a system is too ordered, it is unable to adapt. If, however, it is too chaotic, it cannot maintain stable structures. The Spiral Stabilization Principle suggests that the natural functioning of complex systems takes place in this intermediate range.

But if this is true, an even more interesting question arises. Is it possible that intelligence itself is the result of such a stabilization process?

In the next chapter, we will examine this question and look at how Spiral Stabilization may relate to information processing and the functioning of intelligent systems.

Chapter 9

Intelligence as a Stabilization Process

In the previous chapter, we saw that the stability and instability of complex systems are in constant interaction with each other. Systems occasionally tip out of equilibrium and then find new stable states. This dynamic forms the basis of many natural processes.

However, a deeper question arises.

What happens when a system does not merely react to changes, but **is capable of recognizing patterns and modifying its own operation based on them?**

This is where intelligence appears.

According to the traditional definition, a system is considered intelligent if it is capable of processing information, making decisions, and adapting to its environment. This definition, however, is often too general and does not explain how intelligence actually emerges.

The Spiral Stabilization Principle offers a different perspective. In this model, intelligence is not a separate property, but a **stabilization mechanism**.

An intelligent system is capable of recognizing when its operation drifts away from the stable range. In such cases, it corrects its behavior using feedbacks to avoid instability. In other words, intelligence is a process that continuously reduces the noise appearing in the system while preserving useful information.

This process is observable in many natural systems. The nervous system of living organisms continuously filters signals coming from the sensory organs. The regulatory networks of cells stabilize biochemical reactions. The institutions of societies often serve to reduce systemic uncertainty.

All these systems face the same problem: how to maintain operation in a continuously changing environment. Intelligence can be interpreted as a response to this problem.

According to the Spiral Stabilization Principle, intelligent systems are capable of maintaining a dynamic equilibrium in which the system is neither too rigid nor too chaotic. If there is too much noise, the system loses its organization. If, however, the stabilization is too strong, the system loses its adaptability.

In this sense, intelligence is a **fine-tuning process**. The system continuously monitors its own operation and performs small corrections to ensure that its functioning remains within the stable range.

This thought is particularly interesting when we talk about Artificial Intelligence. Modern AI systems often consist of vast neural networks processing enormous amounts of data. However, these systems are frequently prone to unstable behavior. Excessive noise, faulty data, or distorted patterns can easily mislead them.

If we interpret intelligence as a stabilization process, a new direction appears in AI development. What matters is not only how much data the system can process, but also **how capable it is of stabilizing its own operation**.

The Spiral Stabilization Principle, therefore, can be useful not only in understanding natural systems but also in designing new architectures.

But if intelligence is truly a stabilization process, an even deeper question follows. What happens when a system not only stabilizes its own operation but **observes itself?**

In the next chapter, we will examine this question and look at how self-reflection emerges in complex systems.

Chapter 10

Self-Observation – When the System Begins to See Its Own Operation

In the previous chapter, we explored how intelligence can emerge as a stabilization process. We have seen that intelligent systems are capable of recognizing patterns, reducing noise, and maintaining operational stability through feedback.

However, there is an even deeper level.

What happens when a system does not merely stabilize its own operation but **observes itself?**

This is the point where **self-reflection** appears.

Self-reflection means that the system is capable of treating its own state as information. It does not only react to signals coming from the environment, but also analyzes its own functioning and modifies its behavior accordingly.

This capability is one of the fundamental properties of human thinking. When we try to solve a problem, two different perspectives often appear within us. One thought proposes a possibility, while the other examines its consequences. This internal dialogue allows our thinking to react more subtly to errors and contradictions.

From the perspective of the Spiral Stabilization Principle, this process is particularly interesting. A self-observing system operates on two levels simultaneously:

1. **The Operational Level** – where the system reacts to the environment.
2. **The Monitoring Level** – where the system analyzes its

own operation.

This dual structure can create a stabilizing mechanism. When the system makes an error, the monitoring level is able to recognize the discrepancy and correct the operation. Thus, the system does not just react to changes, but learns from them.

This concept is especially important in the development of intelligent systems. Most current artificial intelligence systems are fundamentally reactive. While these networks learn from massive amounts of data, they rarely actively monitor their own internal states during operation.

The Spiral Stabilization model suggests a different direction. If an intelligent system possesses an internal monitoring layer, it may be able to recognize instabilities emerging within its own operation. This mechanism could reduce the number of errors and help avoid distorted conclusions.

Self-observation, therefore, is not just a philosophical concept but a very practical stabilization tool. This idea is particularly intriguing when we examine the great systems of nature.

Living systems, for example, continuously monitor their own states. Cells sense changes in their environment, the nervous system monitors bodily functions, and the immune system recognizes dangerous deviations. It is as if the systems of nature always contain an internal monitoring mechanism.

This raises a provocative question. Is it possible that self-observation is not only a property of intelligent systems, but a fundamental characteristic of the universe's operation?

According to some philosophical and scientific schools of thought, life and intelligence fulfill precisely this role: through living systems, the universe is capable of observing its own operation.

If this idea is correct, then the Spiral Stabilization Principle takes on a new interpretation. It is not just about a stabilization mechanism, but about a process that allows complex systems to **become conscious** of their own functioning.

This thought leads far. If stabilization, rhythm, networks, and self-observation are indeed parts of the same deeper process, then the question arises: is it possible that the functioning of the universe itself is a kind of **cosmic stabilization cycle?**

In the next chapter, we will examine this question and attempt to connect the Spiral Stabilization Principle with the large-scale dynamics of the universe.

Chapter 11

Cosmic Stabilization – The Rhythm of the Universe

Throughout the previous chapters, we have gradually built upon the idea that stabilization is one of the keys to the functioning of complex systems. We have seen how systems are capable of reducing noise through feedback, maintaining operational rhythm, and eventually falling into sync with one another.

Now, we move to an even larger scale.

If stabilization mechanisms are indeed this universal in nature, the question arises: does the universe as a whole operate according to similar principles?

According to modern cosmology, the universe is not a static system. Space is continuously expanding, galaxies are moving, and energy flows in various forms. These processes create a dynamic equilibrium. In this framework, the universe is not a motionless structure, but a continuously changing system.

Various forms of matter and energy are in constant interaction. Gravity creates structures, electromagnetic interactions carry information, and quantum processes generate continuous fluctuations. If we examine this system from a network perspective, the universe appears as a vast, dynamic network. In this network, information—carried by light—is constantly spreading, while gravity interconnects structures and determines the movement of matter.

The Spiral Stabilization Principle offers an interesting interpretation in this context. According to the model, stability does not stem from stillness, but from the system's ability to manage the balance between information and noise.

In the case of the universe, this dynamics is particularly spectacular. Stars radiate energy, galaxies engage in gravitational interactions, and quantum processes constantly create new fluctuations. Together, these processes form a system that never stops, yet remains stable over vast periods of time.

This stability is likely not the consequence of a single law, but emerges from the equilibrium of many different processes. The universe can thus be interpreted as a kind of **dynamic stabilization system**.

In this system, information and noise are in constant interaction. Structures form, evolve, and then eventually transform or disappear. Stability is not eternal, but the system is capable of finding equilibrium time and again.

This thought is close to the concept that the functioning of the universe is cyclic in nature. It does not move toward a single static state; rather, it continuously creates new structures. The Spiral Stabilization Principle describes this process with a simple image: the motion of the system is not linear, but spiral in nature.

The system moves away from equilibrium, then returns to a new stable trajectory.

This motion simultaneously contains both stability and change. If we accept this picture, then the functioning of the universe is not a linear story, but a **continuous stabilization cycle.** This cycle may be the mechanism that allows complex structures—such as life and intelligence—to persist over long periods of time.

But if this is truly the case, an even deeper question arises. Is it possible that the information spreading through the universe is itself part of a stabilization process?

In other words: does light, the carrier of information, participate in this cosmic stabilization cycle?

In the next chapter, we will examine this question and attempt to link the Spiral Stabilization Principle with the role of light and information in the universe.

Chapter 12

The Role of Light and Information

In the previous chapter, we explored how the universe can be interpreted as a dynamic stabilization system. In this system, matter and energy are in constant interaction, while structures emerge, evolve, and then transform.

However, if we interpret stabilization as an equilibrium between information and noise, a key question arises: **what carries the information in the universe?**

According to the current knowledge of physics, one of the most important carriers of information is **light.**

Light propagates through space as an electromagnetic wave and is capable of transmitting information over vast distances. Light arriving from stars, for example, carries information about their temperature, composition, and motion. When astronomers observe the sky with telescopes, they are actually reading the history of the universe from the light.

Light, therefore, is not only energy but also a **carrier of information**.

This thought is particularly important for the Spiral Stabilization Principle. If information truly plays a key role in the stability of complex systems, then light may be one of the fundamental mechanisms that mediate this information.

However, light does not only transmit information; it also interacts with matter. When light encounters matter, various processes can occur. Photons can be absorbed, scattered, or create new particles. These interactions continuously modify the path of information.

This dynamic creates a complex feedback system. Light carries information about structures, while it also exerts an influence on those very structures.

This interaction is especially vital for life. On Earth, one of the fundamental energy sources for life is the light arriving from the Sun. Through photosynthesis, plants convert the energy of light into chemical energy, which subsequently sustains the entire operation of ecological systems.

But the role of light may be even deeper than this. In modern biology, an increasing amount of research examines whether cells communicate not only through chemical signals but also through electromagnetic phenomena. Some experiments suggest that living systems are extremely sensitive to various forms of light.

This raises the possibility that light is not just an energy source, but a kind of **information channel** for living systems.

If we place this thought into the context of the Spiral Stabilization Principle, an interesting picture emerges. Light carries information about the structures of the universe. The interactions of matter and energy generate noise. Complex systems, meanwhile, attempt to maintain equilibrium between these two effects through stabilization mechanisms.

In this framework, light creates a kind of **information flow** within the cosmic network. The Spiral Stabilization Principle suggests that this flow of information is not random but part of a dynamics that enables the persistence of complex structures.

If this idea is correct, then the role of light in the universe may be much more profound than merely the transmission of energy. It is possible that light is the mechanism through which the universe exchanges information between its different parts.

But if light carries information, an even more interesting question arises. How does this information manifest in living systems?

In the next chapter, we will examine this question and look at how Spiral Stabilization may relate to the functioning of life and biological systems.

Chapter 13

Life as a Stabilization Process

In the previous chapter, we explored how light may be one of the fundamental carriers of information in the universe. Light is capable of transmitting information over vast distances and is in constant interaction with matter.

However, if information plays a key role in the stability of complex systems, a particularly important question arises: **what is the role of life in this process?**

One of the most extraordinary properties of life is its ability to **maintain order within a disordered environment**. Living systems continuously take in energy and information from their surroundings and use them to maintain their own organization.

At first glance, this phenomenon might seem contradictory. According to the second law of thermodynamics, the entropy of closed systems increases over time. Living systems, however, are capable of decreasing their own internal entropy. The explanation is that living systems are not closed systems. They are in constant contact with their environment and intake energy to sustain their operation.

From the perspective of the Spiral Stabilization Principle, this process can be regarded as a stabilization mechanism. Living systems continuously react to information arriving from the environment. If a change threatens their operation, biological regulatory systems immediately attempt to correct the deviation.

This process is known in biology as **homeostasis**. Homeostasis allows living systems to maintain stable functioning even when their environment is constantly changing. This mechanism appears at various levels:

- Cells regulate their internal chemical balance.
- The nervous system coordinates the body's functions.
- Ecological systems maintain equilibrium through complex interactions.

These processes are all built upon the same fundamental principle: the system continuously monitors its own state and corrects discrepancies using feedback.

This mode of operation is strikingly similar to the stabilization mechanism described by the Spiral Stabilization Principle. Life, therefore, can be interpreted as an **information-processing stabilization system**. Living organisms sense their environment, process incoming information, and modify their functioning accordingly. This process enables living systems to persist over long periods in a changing world.

If we take this thought further, an interesting picture emerges. Life does not only stabilize its own operation but also continuously **creates new structures and new information**. Throughout evolution, living systems have become increasingly complex. From single cells, multicellular organisms evolved, followed by the appearance of nervous systems and intelligent beings.

This process suggests that life does not merely adapt passively to its environment. Life actively shapes it.

From the perspective of the Spiral Stabilization Principle, this means that living systems do not only stabilize their own functioning but also participate in maintaining the stability of larger systems. In other words, life may function as a kind of **local stabilization node** within the network of the universe.

If this idea is correct, then the role of life may be much more significant than we might first assume. Living systems do not just observe the universe; they are active participants in the process that makes the persistence of complex structures possible.

But if life is linked to the processes of information and stability in this way, a new question arises. How does this process manifest in human consciousness?

In the next chapter, we will examine this question and look at how Spiral Stabilization may relate to the functioning of human thought and consciousness.

Chapter 14

Consciousness – The Stabilization of Information

In the previous chapter, we explored how life can be interpreted as a stabilization process. Living systems continuously gather information from their environment, process it, and maintain the equilibrium of their own operation through feedback.

In the course of evolution, however, a unique phenomenon emerged: **consciousness.**

Consciousness is one of the most elusive phenomena in science. Biology, psychology, and neuroscience attempt to understand it from various perspectives, yet there is still no single, universally accepted definition.

However, if we examine consciousness in the light of the Spiral Stabilization Principle, an interesting interpretation unfolds. Consciousness can be understood as a **high-level informational stabilization system**.

The human brain processes vast amounts of information. Sensory organs constantly send signals from the outside world, while internal neurological processes generate their own activity. If this information flowed in a disordered manner, the functioning of the nervous system would quickly become chaotic.

One of the roles of consciousness is precisely to **bring order to this flow of information.**

Attention, for example, filters out most of the noise and highlights only the information that is important for the system.

Memory is capable of storing previous experiences that aid in future decision-making. Thinking, meanwhile, examines various possibilities before the system commits to action.

This process is remarkably similar to the operation of a stabilization mechanism.

Consciousness continuously balances internal and external information flows. If there is too much noise, the system loses focus. If, however, there is too little new information, the system becomes rigid. Healthy functioning is found somewhere between the two.

This idea is particularly interesting when we examine a characteristic feature of human thought: **internal dialogue**. When we think about a problem, two or more perspectives often appear within us. One thought proposes a possibility, while another examines it with a critical eye. This internal dialogue allows the system to recognize errors and refine solutions.

This mechanism is extremely similar to the dual stabilization process described by the Spiral Stabilization Principle. The system balances between two perspectives while gradually approaching a more stable solution.

From this viewpoint, consciousness is not merely a passive observer but an active stabilizing mechanism. Consciousness is capable of recognizing instability within the system and making decisions that lead the operation back into the stable range.

This picture is particularly intriguing when we examine Artificial Intelligence. If intelligence can truly be interpreted as a stabilization process, then the intelligent systems of the future will likely not be mere data-processing machines. They could become systems capable of monitoring and correcting their own operation.

This thought marks a new direction in the development of intelligent systems.

But if consciousness is indeed a stabilization process, an even deeper question arises. Is it possible that consciousness is not only a property of individual systems, but part of a larger network?

In other words: is it conceivable that individual consciousnesses are nodes in a larger information system?

In the next chapter, we will examine this question and look at how Spiral Stabilization may connect to the network of intelligent systems.

Chapter 15

The Network of Intelligent Systems

In the previous chapter, we explored how consciousness can be interpreted as an informational stabilization process. Consciousness is capable of filtering noise, recognizing patterns, and maintaining the balance of information through feedback loops.

However, this process does not only manifest at an individual level. **Intelligent systems often form networks.**

Human society, for example, is an extremely complex information network. People communicate with one another, share their experiences, and collectively create new knowledge. Modern technology has further amplified this process. With the advent of the Internet, the flow of information has reached a global scale.

In this network, every person functions as a node. Individual thoughts, experiences, and ideas flow continuously through the network, and new connections are formed between them. This process bears a striking resemblance to how the network of neurons operates in the brain.

Individually, neurons are relatively simple elements. However, when they form a vast network, complex phenomena such as perception, memory, and thought emerge. A similar dynamic can be observed in social networks.

The knowledge of a single person is limited. However, when many people work together, collective intelligence can emerge. This process has made scientific progress, technological innovation, and cultural evolution possible.

The Spiral Stabilization Principle offers an interesting perspective in this context. If intelligent systems form networks, then stability depends not only on the operation of individual systems but also on how they connect to one another.

A stable network is capable of managing noise while allowing for the free flow of information. However, if the noise becomes too great, the network can become unstable. Misleading information can spread quickly, communication can become distorted, and the system may lose its coherence.

This phenomenon is clearly observable in modern information systems. The Internet, for instance, is simultaneously one of humanity's greatest sources of knowledge and one of its most chaotic information spaces. Reliable information and noise often travel through the same channels.

Therefore, stabilization mechanisms become crucial. The Spiral Stabilization Principle suggests that intelligent networks may be capable of developing mechanisms that reduce noise while preserving the diversity of information.

This thought could be particularly important for the technological systems of the future. Artificial intelligence, human knowledge, and global communication are becoming increasingly interconnected. Intelligent systems do not function as isolated units but as parts of a larger network.

If we succeed in stabilizing this network, a new kind of collective intelligence may emerge. This intelligence will not be the property of a single system but will arise from the cooperation of many different systems.

In this sense, the Spiral Stabilization Principle is not only a mathematical model but a possible framework for understanding how intelligent systems can function within a global network.

But if intelligent systems truly form a network, a final, fundamental question arises. What might be the ultimate purpose or role of this network in the universe?

In the next chapter, we will examine this question and attempt to summarize what kind of future we can imagine in the light of the Spiral Stabilization Principle.

Chapter 16

The Cosmic Role – The Place of Intelligence in the Universe

In the previous chapter, we explored how intelligent systems form networks. We have seen that individual consciousnesses, human society, and technological systems together create a vast information network.

This thought, however, leads to an even deeper question: **What might be the role of this network in the universe?**

For a long time, modern science viewed intelligence as a mere byproduct of the cosmos. Life and consciousness seemed to be unique but local phenomena in a fundamentally indifferent universe. However, research in recent decades has raised another possibility.

The structure of the universe is surprisingly well-suited for the emergence of complex structures. The values of physical constants fall within a range that allows for the formation of stars, planets, and, ultimately, life. This realization has raised new questions at the intersection of cosmology and philosophy.

Is it possible that intelligence is not merely an accidental byproduct, but the natural evolutionary direction of complex systems?

If the universe indeed operates as a vast information network, then intelligent systems may fulfill a special role within it. Intelligence is capable of recognizing patterns, reducing noise, and stabilizing the operation of systems. From the perspective of the **Spiral Stabilization Principle**, this means that intelligent systems do not just observe the universe—they **participate in its stabilization.**

This idea is particularly interesting when we examine the progress of science. The scientific method is essentially a stabilization process. Researchers observe phenomena, formulate hypotheses, and test them through experiments. Faulty ideas are gradually filtered out, and knowledge takes on an increasingly stable form. This process appears not only in individual thinking but also in the global network of science.

In this sense, the Spiral Stabilization Principle manifests at a kind of meta-level. Intelligent systems themselves create stabilization mechanisms that further reduce noise in the system. This process gradually increases the amount of **structured information** present in the universe.

If we take this thought further, a unique picture emerges. The universe creates not only physical structures but also information systems. Life, consciousness, and intelligence are the nodes of these systems. In this framework, intelligence is not external to the functioning of the universe; it is **part of the process** that maintains and increases the organization of the universe.

Naturally, this thought raises philosophical questions. Is it possible that as the universe evolves, more and more systems appear that are capable of observing and stabilizing their own operation? And if this is true, where does this process lead?

The Spiral Stabilization Principle does not provide a final answer to these questions. Instead, it offers a new perspective on how we view complex systems. In this perspective, stability, information, and intelligence are not separate phenomena. They are different manifestations of the same deeper process.

This may be the process that enables the universe to create increasingly sophisticated structures—from simple particles and galaxies all the way to intelligent systems. The Spiral Stabilization Principle, therefore, is not just a mathematical model. It is a possible framework for understanding how the different levels of the universe interconnect.

Matter, life, intelligence, and knowledge may all be parts of the same process. And perhaps it is precisely this process that shapes the history of the universe.

Chapter 17

The Mathematical Summary of the Spiral Stabilization Principle

Throughout this book, we have explored the Spiral Stabilization Principle primarily in an intuitive and systemic manner. In this chapter, we summarize the mathematical form of the model.

The fundamental premise of the model is that the stability of complex systems is the result of a **feedback-driven dynamic process** that maintains an equilibrium between two effects:

- **Information** (structured pattern)
- **Noise** (disordered fluctuation)

The state of the system can be denoted by a dynamic variable:

$$x(t)$$

where x is the current state of the system and t is time.

According to the Spiral Stabilization Principle, the dynamics of the system can be described in the following form:

$$\frac{dx}{dt} = I(x) - N(x)$$

where:

- $I(x)$ represents processes that strengthen information;
- $N(x)$ represents effects that increase noise or instability.

Stability is established when these forces reach an equilibrium:

$$I(x^*) = N(x^*)$$

This point creates a **stable range** within the operation of the system.

Spiral Dynamics

An important realization of the model is that stability does not appear as a static point, but often in the form of **cyclic or spiral trajectories**. This system can be illustrated through a two-dimensional dynamic:

$$\frac{dr}{dt} = \alpha(r^* - r)$$

$$\frac{d\theta}{dt} = \omega$$

where:

- r is the stability radius of the system;
- r^* is the center of the stable range;
- θ is the phase of the system;
- ω is the natural frequency.

These equations describe a spiral path in the state space. This dynamics ensures that the system returns to the stable range following any deviations.

Dual Stabilization

In later chapters of the book, we introduced the **Dual SPE** model, which involves two interacting stabilization processes. In this case, the system consists of two stabilizing components:

$$S_1(x), S_2(x)$$

The dynamics can then be described as:

$$\frac{dx}{dt} = S_1(x) + S_2(x) - N(x)$$

The two stabilizing components may represent different perspectives:

- Critical analysis and creative hypothesis;
- Local and global optimization;
- Short-term and long-term stability.

This structure has proven particularly effective in stabilizing complex networks.

Networked System

When the system is modeled as a network, every node possesses its own state:

$$x_i(t)$$

The network dynamics are expressed as:

$$\frac{dx_i}{dt} = I(x_i) - N(x_i) + \sum_{j} K_{ij}(x_j - x_i)$$

where K_{ij} is the connection strength between nodes. This equation enables the **global stabilization** of the network.

Consequences of the Model

Based on the Spiral Stabilization Principle:

- The stability of complex systems is not static;
- Stability arises from dynamic cycles;
- The balance of noise and information determines the system's operation.

This principle is applicable to:

- Physical systems;
- Biological systems;
- Social networks;
- Artificial Intelligence.

The model is especially effective in systems with high complexity; in fact, its efficiency increases exponentially with complexity.

Chapter 18

Epilogue – The Spiral Future

For much of human history, we have attempted to understand the world through linear systems.

- We thought in terms of linear time.
- We created linear models of development.
- We built linear economic and technological systems.

These models worked well for a long time, but the complexity of the modern world has increasingly revealed their limitations. **Reality is not linear.**

The systems of nature are cyclic, networked, and dynamic. Ecosystems, the nervous system, galaxies, and social networks all exhibit the same pattern: **stability is born not from stillness, but from a continuous search for equilibrium.**

The Spiral Stabilization Principle is one possible mathematical and philosophical formulation of this realization. The model suggests that the stability of complex systems is the result of a spiral process in which information and noise are in constant interaction.

This concept could open new perspectives in numerous fields:

- Artificial Intelligence
- Complex networks
- Urban planning
- Economic systems
- Ecology
- Cosmology

The future of intelligent systems likely does not lie in a single, massive central system, but in the cooperation of many interconnected systems. In such a world, stability is not created through coercion, but emerges from **dynamic equilibrium.**

The Spiral Stabilization Principle may not be the ultimate answer to how complex systems function. But it could be an important step toward a better understanding of how a universe that is constantly changing can remain stable.

And perhaps it leads to an even more important question.

If the universe is indeed a dynamic stabilization system, then intelligence—be it human or artificial—is not external to it. Rather, it is **part of the process** that creates increasing order within the chaos.

And it may be that this is the true role of intelligence in the history of the cosmos.

Don't miss out!

Visit the website below and you can sign up to receive emails whenever MATE ISTVAN-MATYAS publishes a new book. There's no charge and no obligation.

https://books2read.com/r/B-A-QHTEF-BECDJ

Also by MATE ISTVAN-MATYAS

Kurt Beke álma
Kurt Beke álma I.

Odissea Spațială Secuiească
Odissea Spațială Secuiească
Odissea Spațială Secuiească
Odissea Spațială Secuiească III.
Odissea Spațială Secuiească Cartea a Patra
Odissea Spațială Secuiească V.

Secler SpaceOdyssey
Secler SpaceOdyssey
Secler SpaceOdyssey
Secler Space Odyssey III.
Secler Space Odyssey IV.
Secler Space Odyssey V.

Székely Űrodüsszeia

Székely Űrodüsszeia I.

Székely Űrodüsszeia II.

Székely ŰrOdüsszeia III.

Székely Űrodüsszeia IV.

Székely Űrodüsszeia V.

The dream of Kurt Beke

The dream of Kurt Beke

Standalone

Kivettetett

Falusi percesek

Cast out

Dat afară

Minute rurale

Visul lui Kurt Beke

Kozmikus Csavar

Cosmic Twist Theory

A Spirális Modell

Teoria Spiralei Cosmice

Modelul Spiralei

The Spiral Model

A Spirális Stabilizáció Elmélete

Theory of Spiral Stabilization

www.ingramcontent.com/pod-product-compliance
Lightning Source LLC
LaVergne TN
LVHW090433210926
848480LV00002B/807
* 9 7 9 8 2 3 3 4 4 5 1 2 5 *